A GREAT AMONG GREATS

ISAMBARD KINGDOM BRUNEL (1806–59) was the foremost engineer in an age of great engineers, an age of excitement, expansion and wealth. The Industrial Revolution was at its height and Britain, its birthplace, was the vibrant hub of a worldwide empire. The power of steam, once confined to mines and mills, had broken free, finding wheels, taking to water, transforming the landscape. If ever there was a time for a man of Brunel's ability, ambition and drive, this was it.

A vivacious, dynamic perfectionist, he drove others hard and himself hardest of all. After helping his father construct the world's first underwater tunnel, he was chosen from more experienced men to design the Clifton Suspension Bridge. From this Bristol connection came the chance to create the Great Western Railway and later to build steamships, the size of which the world had never seen before. The *Great Eastern*, the largest ship ever built before the 20th century, presented challenges and setbacks on a scale to match the size of the ship itself. It was this, his most ambitious project, that undermined Brunel's health and contributed to his premature death at the age of 53.

Much of his work, especially his railway, is still part of Britain's infrastructure today. His visible and splendid legacy makes it easy to think that Brunel's life was throughout one of golden achievement. However, disaster, failure, ridicule and death were never far away – which makes the story of this clever, charismatic, driven man all the more fascinating.

MAIN PICTURE: *Box Tunnel, near Bath. The Great Western Railway was Brunel's finest achievement.*

BIRTH OF AN ENGINEER

SIR MARC BRUNEL
(1769–1849)

A clever and innovative engineer who won his knighthood for completing the Thames Tunnel. Despite the success of his naval blocks, many other ventures were ill-starred. The army suggested he build a boot-making factory, then, in peacetime, refused to accept the boots. A fire in his sawmills made him briefly bankrupt. Despite all the setbacks, Marc Brunel was a gentle and patient father – more so than his son – and spent many hours with the young Isambard. In later life the two remained close and he took great pride in his son's achievements.

AS A MAN, Isambard Kingdom Brunel was on one hand self-assured, determined and resilient; on the other, warm, considerate and full of mischievous good humour. The foundation for this potent combination of qualities lay in his secure and happy childhood.

He was born at Portsea, Hampshire, on 9 April 1806. His mother, Sophia (née Kingdom,) was the daughter of a Plymouth naval contractor, his father, Marc Isambard Brunel, a French engineer. A royalist sympathizer in a new republic, Marc had left for the USA in 1793, six years later coming to Britain with an idea for mass-producing the pulley blocks that sailing ships used to haul up their rigging.

From an early age Marc taught his son the skills and knowledge an engineer would need, among them geometry, drawing techniques and, above all, powers of observation.

At the time, France had higher education in engineering where England did not. So, after boarding school in Hove, Isambard went first to college in Caen and Paris, then worked as an apprentice to Louis Breguet, perhaps the world's finest maker of timepieces and scientific instruments. The young Brunel proved himself a lively, intelligent pupil, often high-spirited and playful, yet inventive and keen to learn where work was concerned. It was from Breguet more than anyone that he learned the importance of detail.

Being English by birth, Isambard was not eligible to go to École Polytechnique, so in 1822 returned to work in his father's small London office. He also spent many hours observing and assisting at the Lambeth works of Henry Maudslay, a superb craftsman and brilliant designer of machines.

Marc, after a succession of misfortunes – he was briefly jailed for bankruptcy in 1821 – nevertheless managed to provide his son with an excellent education for a future engineer. Isambard proved himself a worthy recipient, combining imagination as an artist and ingenuity as a craftsman with a dogged persistence and a rare capacity for hard work. He was to need these qualities in large measure when his father took on the task of building the world's first underwater tunnel.

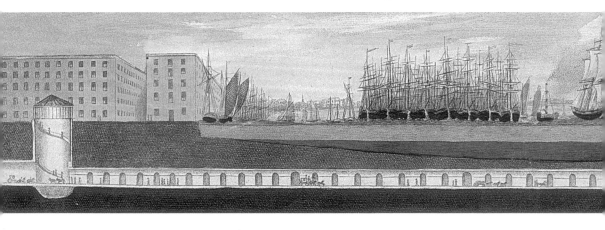

ABOVE: *An idealized 1835 impression of the Thames Tunnel, which finally opened in 1843.*

ABOVE: *The Thames Tunnel was an exacting but valuable training for Brunel, calling for skilled leadership, administration and public relations.*

Tunnelling beneath the Thames had been tried once before. Richard Trevithick, pioneer of steam traction, had overcome repeated floodings to reach the Limehouse shore, but yet another collapse of quicksand in January 1808 had put an end to the project.

Nevertheless, determined backers were impressed by Marc Brunel's tunnelling shield (see panel) and in 1824 appointed him to build a two-way road tunnel from Rotherhithe to Wapping, a few hundred metres west of the previous attempt.

Marc Brunel's tunnelling shield was based on the action of *Teredo navalis*, a worm that ate ships' timbers. It was 6.5 metres (21 feet) high with 12 of the sections pictured above (six for each arch). In each of the 36 'cells' a tunneller excavated one small section only at a time by removing a horizontal board. Behind the diggers, bricklayers consolidated the excavated surface. When all was secure above and behind, the shield was moved forward.

EARLY DISASTER

MARC BRUNEL'S plan for the Thames Tunnel was to build at a shallower level than Richard Trevithick had done, only 4.3 metres (14 feet) below the deepest point of the Thames. Here, geologists assured him, was a layer of clay.

A shaft was ingeniously sunk in soft earth by building a brick tower, reinforced with iron. When undercut by digging, it sank into the ground like a giant pastry-cutter. However, when work on the tunnel proper began, problems soon arose. The geologists had got it wrong; the clay was intermittent and water kept seeping in – black, putrid, stinking water, for in those days the Thames was virtually a sewer. As many as 40 men worked in this to keep the flooding under control. Progress was slow and funding inadequate. The directors therefore sought to cut corners, rejecting a driftway to drain water from the tunnel, and insisting that workmen excavate a larger area at one time than Marc felt was safe. Disaster was inevitable, but disease struck first. Some men were struck blind; one engineer became ill and resigned; another died; Marc Brunel contracted pleurisy.

So it was that Isambard, barely 20 years old, took over as engineer-in-charge, spending up to 36 hours at a time in the tunnel, catnapping when he could. Despite his youth and small stature – only 1.6 metres (5ft 3ins) – Brunel's toughness and dedication earned him respect. Courage was soon added to this list. In May 1827, the collapse which both Brunels had feared happened. Water-borne debris swept through the tunnel. Isambard, never one to shirk danger, let himself down into the shaft to rescue an old winch man. Fortunately no one died.

Repairs were effected and in November Brunel, with trademark showmanship, announced this to the world by holding a candlelit banquet in the tunnel.

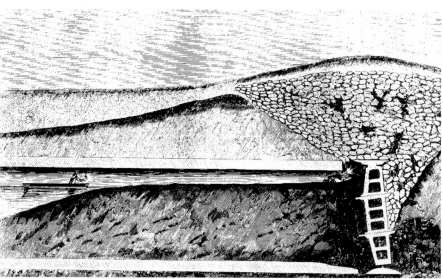

ABOVE AND TOP: As a young man, Brunel showed little fear of danger. To examine the extent of the 1827 tunnel collapse, he first had himself lowered to the river bed in a diving bell (top), then later crawled into the flooded workings from a punt. A man died ten days later undertaking a similar excursion.

But, less than two months later, at high tide in the early hours of a January morning, disaster struck again. Once more a tide of foul water rushed in. Timber from working platforms fell on Brunel, who was swept, barely conscious, up and out of the shaft, to be miraculously grabbed by an engineer, seconds before he would have gone back down for ever. Six men did die that night, and the project was abandoned for seven years. Brunel would never work on the tunnel again.

His internal injuries were severe and he was sent to Brighton to convalesce. However, his characteristic restlessness caused him to overdo things, and he returned to London. After several weeks' sojourn at his parents' home he went, perhaps for further convalescence, to Clifton, a fashionable suburb of Bristol overlooking the spectacular Avon gorge. It was here that Brunel was to meet the wealthy merchants who would shape his working life.

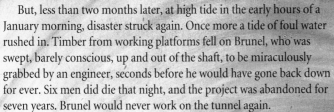

HENRY MAUDSLAY
(1771–1831)

A brilliant mechanic and craftsman, Maudslay was the originator of the modern machine shop, designing lathes, planes etc. far more accurate than anything before. His company, Maudslay, Son & Field, the leading engineering firm of the time, was an integral part of many Brunel projects, from the Thames tunnelling shield to the engines of the *Great Western*.

Hundreds of visitors paid each day to view the tunnel, which both Brunels thought highly dangerous. Nevertheless, in late 1827 Isambard held a candlelit dinner to signal the restart of work.

OPPORTUNITY KNOCKS

DURING HIS STAY in Bristol, Brunel no doubt spent time walking the heights overlooking the nearby Avon gorge. One can imagine his excited reaction when, in 1829, a competition was announced to design a bridge spanning the gorge.

In assisting his father, Brunel already had some experience of suspension bridges. To add to this knowledge he made careful studies of Telford's Menai bridge, inspected other bridges and took advice from trusted friends such as Maudslay. He then submitted four beautifully painted designs for different gorge locations. These took their place amongst the efforts of others, many of them weird and wonderful.

Thomas Telford was invited to judge the designs; the veteran engineer duly rejected them all – Brunel's because the length of span, he said, offered inadequate wind resistance. Telford himself then produced a curious design which involved two ornate Gothic piers rising from river level. Brunel wrote to the bridge committee condemning its structural timidity, but it was mainly public opposition to its appearance which saw it defeated.

In 1830 a further competition was held, Brunel's entry this time coming second because the bridge committee's advisors had reservations about the strength of his innovative chains, anchorage and suspension rods. It was here that the young man's eloquence and charisma came to the fore. Arming himself with copious drawings, in a matter of hours he succeeded in convincing the adjudicators that his plan was indeed technically sound. Unanimously they adopted his Egyptian design (see picture below).

TOP: Brunel's 'Giant's Hole' design for the Clifton bridge was reputedly his favourite.

ABOVE: Stuck high above the gorge on one traverse beneath the iron cable slung between the piers, Brunel climbed out of the basket to free the jammed pulley.

LEFT: The design that Brunel called his 'Egyptian thing' was the one which won him the commission.

Brunel's euphoria was not to last. Although a token start to the work was made in June 1831, subscriptions from backers fell far short of the projected £45,000 cost. Worse followed, for later that year political rioting in Bristol put a stop to any further finance.

For the next three years the lack of progress reflected Brunel's career in general. However, in 1836 financial confidence returned and work began again in earnest, proceeding in fits and starts until 1843, by which time the piers, approach roads and anchorages were ready and the ironwork ordered. Just one problem remained – there was no money left, and a further £30,000 was needed.

In 1851 chains and plant were sold off to pay creditors and there was even talk of demolishing the two piers. For nine more years they stood as lonely monuments to a long-held dream. Ironically it was Brunel's death in 1859 that was to inspire the bridge's completion. The Institution of Civil Engineers required a suitable monument for him, and by happy coincidence the Hungerford footbridge in London, designed by Brunel, was to make way for a new Charing Cross railway bridge; its chains would therefore require disposal. Engineers and Clifton trustees met and the Hungerford chains were duly purchased. Construction was rapid, and the Clifton Suspension Bridge was ceremonially opened in December 1864.

BELOW: *For 20 years, the stone bridge piers stood isolated on their cliff tops. Only in 1864 were the chains (re-used from the Hungerford footbridge, another Brunel project) and deck installed.*

This painting by Brunel's brother-in-law, John Horsley, shows the engineer with his plans for the Great Western Railway.

CASTLES IN THE AIR

BRUNEL'S SUCCESS in the Clifton competition was a beacon in an otherwise dark landscape. While Thomas Telford had built over 1,600kms (1,000 miles) of road and 1,200 bridges, and George and Robert Stephenson were about to complete their Liverpool–Manchester railway, Brunel had no completed projects to show the world what he could do. 'What will become of me?' he asked in his diary of 1827.

That same diary reveals his '*chateaux d'Espagne*' or 'castles in the air' as he called them – elaborate visions of how he might achieve fame and fortune. Among these were a new London Bridge, a fleet of ships to storm Algiers, new tunnels at Liverpool and Gravesend, and a fine house to reflect the wealth he would achieve.

One earnest ambition was to create a 'gaz engine', a step up from steam power, harnessing other gases than water vapour; but ten years of often dangerous experiment were to end in weary failure.

However, a major aspect of Brunel's greatness was his refusal to give in to setbacks. 'No job too small' was his philosophy, and for several years he set an amazing pace, travelling by coach and carriage the length and breadth of the country, seeking commissions wherever and whatever they might be – drainage works in Essex, a dock scheme in Sunderland, an observatory in Kensington.

Amid this hectic schedule, Brunel always made opportunities to visit sites of architectural or engineering interest: bridges, canals, cathedrals and the latest development … railways.

By 1831, the enormous potential of rail travel preoccupied many engineers. It certainly excited Brunel. In that year he travelled on the newly-opened Liverpool–Manchester railway. His spidery drawings executed while on the train provide an enduring record of how shaky and uncomfortable he found the journey. There and then he resolved, given the chance, to create something faster and smoother.

The opportunity, however, took some time to arrive. In 1830 he narrowly missed being appointed to build the Newcastle–Carlisle railway. Later that year discussions about a proposed Bristol–Birmingham railway foundered through lack of financial support. Yet it was through his Bristol connections that Brunel got the commission that would make his name. A smaller job came first, though.

For many years the city's docks had been allowed to become choked with silt. Bristol's mud was losing trade to Liverpool's deep water and in 1832 Brunel was asked to recommend improvements. Despite some delays, his plans were largely implemented, cementing his relationship with the City Fathers.

Unknown to Brunel, a group of Bristolians had met in autumn 1832 to discuss the building of a railway to London. By the following February the idea had gained momentum. The young engineer learned that a new and powerful railway committee needed someone to survey the line. Brunel felt that his hour had come.

ROBERT STEPHENSON
(1803–59)

Robert Stephenson was the son of George Stephenson, founder of the railway age. Having managed his father's locomotive works, he branched out to build many railways and famous bridges. Although rivals in public, he and Brunel were lifelong friends, offering each other advice and support whenever possible. Stephenson particularly backed Brunel in his battle to build the Great Western Railway, and was regularly at his side during the traumatic building of the *Great Eastern*.

ABOVE: *Travelling on George and Robert Stephenson's Liverpool–Manchester railway made Brunel aspire to building something faster and more comfortable.*

A STAR IS BORN

ABOVE: *Brunel in his ground-floor office in Duke Street, near Whitehall, London. He and his wife, Mary, lived in great style in the rest of the house.*

BRISTOL'S RAILWAY CONSORTIUM proposed that four competing engineers should make a preliminary survey of the line. Whoever gave the lowest estimate for building the line would get the job.

We know from his diaries how badly Brunel wanted the commission, yet, with typical bravura, he declined to join the competition, declaring that he would build the best 'road', not the cheapest. The Brunelian blend of high principle and chutzpah won the day and on 6 March 1833 he was confirmed as engineer to the Bristol Railway.

The 26-year-old Brunel had until May to complete the preliminary survey, and embarked on a punishing schedule that was to become a way of life. By day he shuttled back and forth along the proposed route in coach or on horseback, constantly weighing up how to minimize the engineering needed and thus save time and money. To make these forays easier he designed the 'Flying Hearse', a carriage that would accommodate his plans and instruments, but also afford him the chance of an hour or two's sleep. By night, in many locations, he wrote reports and drew plans, not just for the railway, but for other works he was still engaged upon. In September a detailed survey of the newly-named Great Western Railway was commissioned, the directors forming two committees, in Bristol and in London.

Building a railway was more than just an engineering matter. Money had to be raised, owners persuaded to part with their land. The third, and most daunting, challenge for the directors was in securing the approval of Parliament. It was here that the opposition – obdurate landowners, stagecoach and canal proprietors, promoters of rival railways – massed in strength, throwing up every conceivable argument why the line should not be built. The Provost of Eton even maintained that it would threaten the morals and discipline of his pupils.

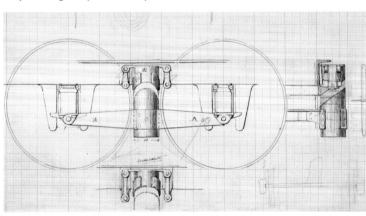

ABOVE: *Brunel's notebooks reveal his eye for detail as well as his sense of the larger vision.*

It was in these more abstract areas that Brunel showed another aspect of his greatness. In the drawing rooms of the wealthy he proved patient and tactful; in public assembly rooms he was eloquent and persuasive. But in the committee rooms of Westminster he was all this and more. In 1834 the House of Lords had thrown out a Bill to build limited stretches of the line, but in 1835 the complete scheme came once more to the House of Commons committee stage.

Ruthlessly cross-examined for 11 days, Brunel gave an inspired performance, totally belying his passionate nature. In the face of ignorance and provocation, he was unflustered in manner and lucid in explanation, calmly and logically refuting every counter-argument to the railway scheme. Above all he showed an absolute mastery of his facts. The young engineer emerged as a star. Moreover, his Great Western Railway was going to be built!

CHARLES SAUNDERS (1795–1864)

One of Brunel's staunchest friends and allies in the debates which he invariably provoked, Saunders was secretary to the Great Western Railway's London committee, later becoming its chief administrator. A quiet, hard-working man, Saunders was tireless in his efforts to find and cajole potential backers for the railway. Like Brunel, he was also very persuasive in parliamentary committee.

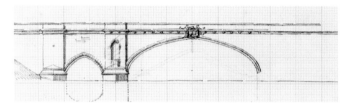

ABOVE: *Brunel's sketch for a railway bridge over the Avon at Temple Meads, Bristol, shows his considerable artistic flair.*

BELOW: *A map from the 1920s of Brunel's grand vision. The London–Bristol route was conceived as merely the spine of a much larger network.*

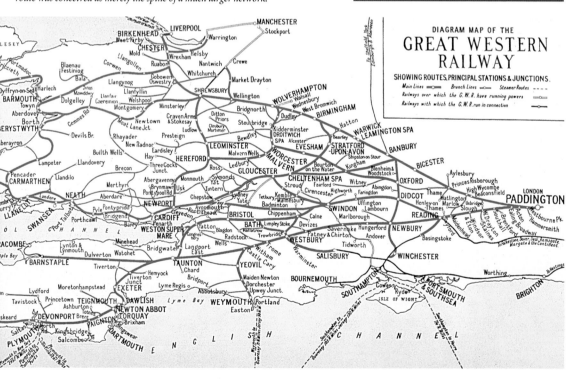

BRUNEL'S BILLIARD TABLE

THE YEAR 1835 was an important one in Brunel's life, as many of his plans were coming to fruition. Besides the Great Western Railway, he had several smaller projects to oversee, working almost round the clock as a matter of routine. Such was the pace he set himself that his diary, previously kept so faithfully, became sketchy and irregular. Yet whatever the pressure, whatever the time of day, Brunel never seemed to lose his buoyant good humour. Another outstanding characteristic was his ability to nurture a grand vision yet retain an amazing command of detail.

His vision for the GWR was of a vast network of branches spreading from the main London–Bristol trunk route, but Brunel also knew every inch of the line, a fact borne out by the many detailed drawings in his notebooks and his command of facts in parliamentary committee.

While other railways were being built chiefly to move freight, Brunel saw the GWR's prime purpose as carrying passengers comfortably and at speed. So he kept gradients low – hence the nickname 'Brunel's billiard table' – and made a late decision to adopt a 7-feet (broad gauge) track compared with the 4 feet 8½ inch gauge used elsewhere, calculating that this would give greater stability at speed.

MARY BRUNEL
(1813–81)

Brunel's wife, Mary Horsley, came from a highly cultured Kensington family. Nicknamed 'the Duchess of Kensington' by her younger sisters, she was coolly beautiful, her level-headed personality being a perfect foil for Brunel's restless dynamism. She seemed perfectly content to remain in their opulent Duke Street home, raising children and presiding over the many drawing room theatrical and musical events that went on there.

Having established an office of his own in Duke Street, near Whitehall in Westminster, Brunel had to find and train assistants to his own high standards, not an easy task when all the experienced railway men were at work elsewhere. An archetypal Victorian, driven by his conscience, he led from the front. He would brook no interference from his directors on engineering matters and could be scathingly intolerant of inferior work from those he employed.

It is not surprising therefore that progress on the railway was rapid. In all weathers, gangs of navvies toiled for up to 16 hours a day, aided only by horses, and huge wheelbarrows which could carry up to 200kg (4cwt) of earth per load. Whether from respect for Brunel's example, the magnetism of his personality or fear of his wrath, there were no major labour disputes. The whole GWR project was to take only five and a half years.

With two sisters and lively personal charm, Brunel was at ease with women, yet until now his only constant companions had been his work and his ever-present cigar. However, since first meeting Mary Horsley in 1832, he had been attracted to her and now felt able to offer her the security he deemed essential. They were married in July 1836, setting up house above and behind Brunel's office. Here they lived in fashionable splendour. In succeeding years they would have three children: Isambard junior, Henry Marc and Florence.

Although his work preoccupied him and regularly took him away, Brunel was a loving father with a terrific sense of mischief. Indeed, in performing a trick for his children, he nearly lost his life after swallowing a coin.

ABOVE: *In appalling conditions, Brunel supervised 1,200 men and nearly 200 horses in excavating Sonning Cutting near Reading.*

LEFT: *One of Daniel Gooch's locomotives emerges from a tunnel near Bristol.*

ABOVE: *The railway 'boom' owed much to the industry and hardiness of 'navvies'.*

TRIUMPH AND DISASTER

Gooch's superb locomotives, from *Firefly* to *Iron Duke*, enabled Brunel to fulfil his vision of high-speed passenger transport. A young ex-employee of Robert Stephenson, he admired the broad gauge and was appointed Chief Locomotive Assistant to the GWR, becoming Brunel's staunch friend. It was Gooch who used the *Great Eastern* to lay the Atlantic cable, afterwards rejoining the GWR as Chairman.

THE FIRST STRETCH of the GWR, from Paddington to Maidenhead, opened in June 1838. It was a huge disappointment. The locomotives, which Brunel himself had designed, proved unreliable and short of power, and the ride as uncomfortable for passengers as railways elsewhere. Brunel was heavily criticized, one vocal group of northern shareholders demanding his resignation.

He was rescued by Daniel Gooch (see panel). Not only did Gooch coax better performance from the existing locomotives, but he also brought in Stephenson's *North Star* and *Morning Star*, the first of the 'broad gauge flyers'. Gooch's own class, *Firefly* and others, followed – bigger, faster locos which achieved some amazing journey times.

Construction, too, was advancing rapidly. In 1840 the line reached from London to Reading and the Bristol–Bath section opened. Little-known Swindon was chosen for the main depot and workshop.

Brunel shuttled to and fro, working 20-hour days. His main achievements included the controversial low-arched bridge over the Thames at Maidenhead and Sonning Cutting, 3km (2 miles) long and 18m (60 feet) deep. Under Brunel's personal direction, hundreds of men and horses had battled against storms and mud to complete it.

However, the biggest triumph was Box Tunnel near Bath, at almost 3km (2 miles) the longest railway tunnel built at that time. It took nearly five years and cost 100 lives to complete, towards the end involving an astounding 4,000 men and 300 horses toiling by candlelight. The railway opened in June 1841 at a cost of £6.5 million, every detail of its elegant landscape having had its origin in Brunel's imagination. From the outset it was the world's fastest line.

Already, the railway had pushed south-west from Bristol, but not without setbacks. The South Devon Railway was to be a broad gauge extension

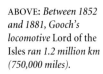

ABOVE: *Between 1852 and 1881, Gooch's locomotive* Lord of the Isles *ran 1.2 million km (750,000 miles).*

RIGHT: *A comical look at the chaos caused by having to transfer from one gauge to another.*

LEFT AND BELOW: *The Royal Albert Bridge over the River Tamar at Saltash, today and, below, during construction. Having been aligned, each truss was jacked up to its final position over some days, as masons built the stonework beneath.*

of the GWR from Exeter to Plymouth. Brunel again sought high-speed running, but the route ahead was hilly and tortuous. The answer, he felt, lay in a new atmospheric system. Stationary steam engines pumped air out of a tube which ran between the rails. The partial vacuum created in the pipe propelled along a piston, which was attached to the train.

Expensive to build, the first section of railway, opened in 1846, proved hopelessly unreliable, minor breakdowns bringing the whole line to a halt. Within 18 months the leather seal on the pipe was found to be perishing and was prohibitively costly to replace. The atmospheric system was scrapped.

To reach Cornwall, the Tamar estuary had to be crossed. The resulting Royal Albert Bridge at Saltash, was one of Brunel's finest railway achievements. Naval height requirements ruled out a timber bridge, so the engineer used two massive bow-shaped tubes of wrought iron to form 1,000-tonne elliptical trusses supported by a central pier. On 1 September 1857, after meticulous planning, the first truss was manoeuvred by barges into alignment with the piers, prior to later jacking into position. The procedure, watched by thousands of awestruck onlookers, was theatrically directed by the great man himself.

ABOVE: *The only surviving engine house from Brunel's doomed atmospheric railway is at Starcross, on the estuary of the River Exe in Devon.*

THE GREAT BRITAIN

BRISTOL

Brought back to the UK in 1970, ss Great Britain is being painstakingly restored in the Bristol dock where she was built.

STEAMSHIPS ACROSS THE OCEAN

IT WAS GENERALLY CONSIDERED that no steam vessel could hold enough coal for ocean voyages. Brunel, however, realized that when a ship moves forward, the counteracting resistance of the water – the factor which determines the amount of fuel needed – is dictated not by its volume, but by the area of its cross-section. Therefore, the bigger the vessel, the smaller the proportion of its space needed for coal.

In 1835, amid plans for the new railway, he suggested building a transatlantic steamship, and in July 1836 construction of the ss *Great Western* began in Bristol. This timber-hulled paddle steamer would be the world's largest steamship. Brunel's insistence on longitudinal strength entailed a massive oak construction which in turn required two huge Maudslay engines, fitted in London in the winter of 1837–38.

However, Liverpool men, with new steamships on order, also wanted the Atlantic passenger traffic and the lucrative mail contract. To steal Bristol's thunder, on 28 March 1838 they dispatched *Sirius*, an Irish packet boat so unsuited to the Atlantic that its crew were eventually reduced to stripping the interior woodwork to keep the boilers going as the ship neared New York.

The *Great Western*, however, had problems of its own. On 31 March, two hours into her maiden voyage, the boiler lagging caught fire, and with it the deck above. Brunel, amid the smoke, fell headlong into the boiler room, a fall miraculously broken by Claxton, who then saved the engineer from drowning in the accumulated water below.

Despite the setback, *Great Western* reached New York only hours behind *Sirius* and with coal to spare. In the next eight years, she completed 67 further crossings. Liverpool, however, won the mail contract.

ABOVE: *ss* Great Britain, *launched in July 1843.*

In July 1839, too late in the day, the keel of a new Bristol ship, the ss *Great Britain*, was laid. This was a revolutionary vessel: twice the size of its Liverpool rivals, it was the first large iron-hulled ship and the first passenger vessel to have a screw propeller, a late decision based on Brunel's successful *Rattler* experiments for the Royal Navy (see picture left).

Launched in July 1843, *Great Britain* was only able to depart from Bristol late the following year, on an abnormally high tide. Even then Brunel had desperately to remove masonry to allow her passage. On her second Atlantic run she broke a propeller. On her fifth voyage she ran aground at Dundrum Bay, Northern Ireland, where she lay for some months, as a furious Brunel put it, 'like a useless saucepan'. Thanks to him, though, she was structurally very strong and was eventually saved. The company, however, foundered. Both their vessels were sold off, the *Great Britain* being converted to sail in the 1880s. An onboard fire in 1886 saw her beached in the Falklands until her rescue in 1970. She is now proudly berthed in the dock where she was built.

ABOVE: *Screw versus paddle. In Brunel's tug-of-war, the specially-built, screw-propelled HMS* Rattler *(left) convincingly beat the* Alecto.

ABOVE: *ss* Great Britain *aground in Dundrum Bay. Only her massive longitudinal strength saved her from destruction.*

THE BIGGEST SHIP IN THE WORLD

DESPITE HIS VIVACITY AND DRIVE, Brunel was not a healthy man. Tireless work, cigars and mishaps had all taken their toll. Added to this was disillusion: the Avon gorge was still unbridged, his steamships had been sold off and the railway bubble had burst, putting many of his assistants out of work. By 1850, retirement to Devon looked increasingly attractive.

Two projects, however, rekindled Brunel's imagination: an involvement in planning the Great Exhibition and a commission to supervise the building of two steamships for the Australia run, which would be via a stop in South Africa to take on coal. How much more economical, Brunel thought, would be a vessel large enough to go there and back without coaling.

John Scott Russell, a marine engineer who had worked with Brunel on the Great Exhibition, agreed to collaborate with Brunel in putting this idea to the Eastern Steam Navigation Company who, having recently failed to win the Australian mail contract, were eager for a grand scheme to raise their profile. They backed the pair in building an iron ship six times bigger than anything ever seen before, a ship needing screw **and** paddle propulsion.

Brunel was deeply committed to the project, investing his own money and persuading others to do likewise. Work began in February 1854, the massive hull soon dwarfing Russell's yard at the remote, marshy tip of the Isle of Dogs. But only a year later, Russell had run

ABOVE: *Brunel, looking far older than his years, in an engraving from an 1854 photograph.*

The Great Eastern in construction at Millwall.

into financial trouble. A disastrous fire found him under-insured, besides which the sheer size of the *Leviathan* (as the ship was first called) was making unanticipated demands on the manufacturing infrastructure. The bank granted further loan payments only on condition that Brunel certified that the work had been done.

Relationships deteriorated. Brunel and Russell disagreed over how the ship was to be launched, Russell wanting a conventional stern-first launch, Brunel insisting on a controlled sideways launch, costly but far safer. Progress slowed. Letters, once cordial, became more and more frosty as a blow-by-blow battle developed. To work out a launch procedure, Brunel needed figures from Russell, but got no satisfactory response. Brunel then threatened to withhold certification. Russell in turn complained that Brunel was continually adding extra work to the job. Brunel countered that Russell's financial demands did not match the progress made. The last straw for Brunel came when the yard began work on other ships, one directly in *Great Eastern*'s launch path. Reluctantly he reported the problems to the directors.

Investigation revealed that Russell had been hiding the gravity of the situation, claiming for materials not received and mortgaging the yard to the hilt. Brunel, in trying to save both Russell and his ship, had authorized more money than results merited. Nearly £300,000 had been spent with only a quarter of the work on the hull done. In February 1856, the bank pulled the plug on the project and work stopped.

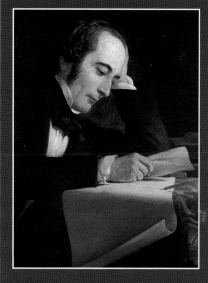

JOHN SCOTT RUSSELL (1808–82)

There is no doubt that Russell was a brilliant and innovative marine engineer. It is hard to assess how much of his action (or inaction) concerning the *Great Eastern* was an understandable response to an over-ambitious scheme. He was certainly not the manager that Brunel was; financially, too, he was out of his depth, and appears to have tried to disguise this in various ways which appalled Brunel's high sense of probity.

A SUPERHUMAN TASK

AFTER MUCH WRANGLING, work on the *Great Eastern* began again late in May 1856, against a strict schedule imposed by exasperated creditors. Brunel agreed, reluctantly, to attempt a launch on the tide of 3 November 1857, working day and night to prepare. The 12,000-tonne hull (the largest weight ever moved by man), resting on two huge cradles, would be pulled by winches and, if necessary, pushed by hydraulic rams, 75 metres (240 feet) to the water. Brunel knew from his experiments that the ship might well not move at first. In any case, the launch would be unspectacular and very tricky. He requested an empty yard on the day. The company sold 3,000 tickets!

The 'event' was a sad parody of the Saltash bridge event. Rain fell on a dejected crowd. The huge ship gave an unnerving lurch, fatally injuring a workman, but then stayed where it was. Newspapers blamed Brunel for this 'failure'.

ABOVE: *Brunel knew that launching the* Great Eastern *would be slow and unspectacular. Yet, against his wishes, the owners invited thousands to watch.*